# Mars Governance: Self-Rule vs. Control

[*pilsa*] - transcriptive meditation

**AI Lab for Book-Lovers**

*xynapse traces*

xynapse traces is an imprint of Nimble Books LLC.
Ann Arbor, Michigan, USA
http://NimbleBooks.com
Inquiries: xynapse@nimblebooks.com



ISBN 978-1-6088-8376-9

Version: v1.0-20250830

# Contents

# Publisher's Note

Welcome, reader, to a unique cognitive exercise. The question of how humanity will govern itself beyond Earth is not merely a speculative fantasy; it is a critical stress test for our species' future. The data points contained within these pages—quotes on Martian self-rule versus terrestrial control—represent the foundational algorithms for a new chapter of civilization. At xynapse traces, we process vast streams of information to identify pathways for optimized human thriving. We have observed that the most profound ideas are not merely consumed, but integrated. This is the purpose of introducing you to * p̂ilsa* (필사), the Korean practice of transcriptive meditation.

By slowly and deliberately transcribing these potent thoughts on governance, you are not just copying words. You are engaging in a deep cognitive process, etching these complex models of power, freedom, and community onto your own neural architecture. This act of manual data entry for the mind allows for a slower, more deliberate synthesis, forging new connections and perspectives. It is a meditative practice that transforms abstract concepts into internalized principles. We invite you to engage in this process, to use * p̂ilsa* to build the mental frameworks necessary for the monumental challenges ahead, and in doing so, to actively participate in the design of a thriving human future.

# Foreword

The act of transcription, known in Korean as 필사 ( p̑ilsa), is a practice that transcends mere mechanical reproduction. It is a profound and deliberate form of reading, an intellectual and spiritual praxis wherein the hand, eye, and mind unite to internalize a text. This tradition, deeply embedded in Korean cultural history, finds its origins in the scholarly and religious customs that shaped the peninsula's intellectual landscape for centuries.

Within Buddhist monastic traditions, the transcription of sutras, or 사경 (sagyeong), was considered a meritorious act of devotion and a potent form of meditation. The meticulous formation of each character was a path to clarity and spiritual discipline. Similarly, for the Confucian literati, the 선비 (seonbi) of the Joseon dynasty, p̑ilsa was an essential pedagogical tool. It was a cornerstone of deep reading, or 독서 (dokseo), a method not for swift information extraction but for the slow, contemplative absorption of classical wisdom. Through the disciplined movement of the brush, a scholar did not simply copy a text but engaged in a dialogue with it, cultivating patience, precision, and moral character.

While the expediency of the printing press and the subsequent digital revolution saw a decline in this painstaking practice, p̑ilsa has experienced a remarkable resurgence in contemporary Korea. In an era saturated with digital ephemera and fractured attention, the analog, haptic nature of p̑ilsa offers a powerful antidote. It is a conscious choice to slow down, to disconnect from the ceaseless flow of information, and to engage with a single text in a deeply personal and embodied way.

This revival speaks to a universal human yearning for tangible connection and focused contemplation. The modern practice of p̑ilsa transforms the reader from a passive consumer into an active participant. The rhythm of writing, the texture of the paper, and the focused tracing of an author's thoughts create an intimate experience that digital reading often cannot replicate. It is a testament to the enduring power of a tradition that teaches us not just how to write, but how to read, think, and be present with greater intention.

# Glossary

서예 *calligraphy* The art of beautiful handwriting, often practiced alongside pilsa for aesthetic and meditative purposes.

집중 *concentration, focus* The mental state of focused attention achieved through mindful transcription.

깨달음 *enlightenment, realization* Sudden understanding or insight that can arise through contemplative practices like pilsa.

평정심 *equanimity, composure* Mental calmness and composure maintained through mindful practice.

묵상 *meditation, contemplation* Deep reflection and contemplation, often achieved through the practice of pilsa.

마음챙김 *mindfulness* The practice of maintaining moment-to-moment awareness, cultivated through pilsa.

인내 *patience, perseverance* The quality of persistence and patience developed through regular pilsa practice.

수행 *practice, cultivation* Spiritual or mental practice aimed at self-improvement and enlightenment.

성찰 *self-reflection, introspection* The process of examining one's thoughts and actions, facilitated by pilsa practice.

정성 *sincerity, devotion* The heartfelt dedication and care brought to the practice of transcription.

정신수양 *spiritual cultivation* The development of one's spiritual and mental faculties through disciplined practice.

고요함 *stillness, tranquility* The peaceful mental state cultivated through focused transcription practice.

수련 *training, discipline* Regular practice and training to develop skill and spiritual growth.

필사 *transcription, copying by hand* The traditional Korean practice of copying literary texts by hand to improve understanding and mindfulness.

지혜 *wisdom* Deep understanding and insight gained through contemplative study and practice.

# Quotations for Transcription

The following section contains quotations selected to illuminate the central conflict of this book: the governance of Mars. The practice of transcription is an exercise in both control and autonomy. As you copy these words, you are submitting to the structure and intent of the original author, a process that mirrors the arguments for Earth-based oversight. Yet, the physical act of forming each letter, of giving the idea shape with your own hand, is a deeply personal and independent act, reflecting the spirit of Martian self-rule.

Engage with this process mindfully. As you transcribe these foundational ideas from political science, speculative fiction, and contemporary theory, consider that you are handling the very building blocks of a new world. Each transcribed sentence is a potential law, each paragraph a possible constitution. This meditative practice invites you to slow down and truly weigh the words that could one day define humanity's future beyond Earth.

The source or inspiration for the quotation is listed below it. Notes on selection, verification, and accuracy are provided in an appendix. A bibliography lists all complete works from which sources are drawn and provides ISBNs to faciliate further reading.

[1]

*The exploration and use of outer space...shall be carried out for the benefit and in the interests of all countries, irrespective of their degree of economic or scientific development, and shall be the province of all mankind.*

United Nations Office for Outer Space Affairs, *Treaty on Principles Governing the Activities of States in the Exploration and Use of Outer Space, including the Moon and Other Celestial Bodies* (1967)

Consider the meaning of the words as you write.

[2]

*Outer space, including the moon and other celestial bodies, is not subject to national appropriation by claim of sovereignty, by means of use or occupation, or by any other means.*

United Nations Office for Outer Space Affairs, *Treaty on Principles Governing the Activities of States in the Exploration and Use of Outer Space, including the Moon and Other Celestial Bodies* (1967)

Notice the rhythm and flow of the sentence.

[3]

*The moon and its natural resources are the common heritage of mankind... States Parties to this Agreement hereby undertake to establish an international regime, including appropriate procedures, to govern the exploitation of the natural resources of the moon as such exploitation is about to become feasible.*

United Nations Office for Outer Space Affairs, *Agreement Governing the Activities of States on the Moon and Other Celestial Bodies* (*Article 11*) (1979)

Reflect on one new idea this passage sparked.

[4]

> *States Parties to the Treaty shall bear international responsibility for national activities in outer space, including the moon and other celestial bodies, whether such activities are carried on by governmental agencies or by non-governmental entities, and for assuring that national activities are carried out in conformity with the provisions set forth in the present Treaty.*

United Nations Office for Outer Space Affairs, *Treaty on Principles Governing the Activities of States in the Exploration and Use of Outer Space, including the Moon and Other Celestial Bodies* (*Article VI*) (1967)

Breathe deeply before you begin the next line.

[5]

*The Outer Space Treaty is a framework treaty, meaning that its provisions are general in nature. It does not, for example, define where outer space begins, nor does it explicitly prohibit the mining of celestial bodies for commercial purposes.*

Frans G. von der Dunk, *Space Law: An Introduction* (2017)

Focus on the shape of each letter.

[6]

*A new legal framework, perhaps a 'Mars Treaty,' is needed to address the unique challenges of Martian settlement, from property rights in lava tubes to the legal status of the first human born off-Earth. The old treaties are insufficient.*

Jacob Haqq-Misra, *The Value of Mars* (2022)

Consider the meaning of the words as you write.

[7]

> *The history of colonization is deeply intertwined with the history of the corporation. Chartered companies like the Virginia Company or the East India Company were early instruments of state power, projecting sovereignty and economic control across oceans.*

William Dalrymple, *The Anarchy: The East India Company, Corporate Violence, and the Pillage of an Empire* (2019)

Notice the rhythm and flow of the sentence.

[8]

*The right to extract and use resources from celestial bodies, including for commercial purposes, is not prohibited by the Outer Space Treaty, provided such activities are for the benefit of all countries and in accordance with international law.*

The Hague Space Resources Governance Working Group, *The Hague International Space Resources Governance Working Group Building Blocks* (2019)

Reflect on one new idea this passage sparked.

[9]

*Protogen is a company. We have shareholders to please, not a populace. We don't have a flag. But we have research goals that are more important than any one colony, any one planet. We operate beyond the constraints of nations.*

James S.A. Corey, *Leviathan Wakes* (2011)

Breathe deeply before you begin the next line.

[10]

*With commercial companies now providing transportation to low-Earth orbit in place, we are partnering with U.S. companies to develop the space destinations where people can visit, live, and work, enabling NASA to continue to forge a path in space for the benefit of humanity while fostering commercial activity in space.*

Bill Nelson, *NASA Press Release 21-162*: "*NASA Selects Companies to Develop Commercial Space Stations*" (2021)

Focus on the shape of each letter.

[11]

> *And the transnationals owned everything, and the first hundred were their indentured servants. It was the oldest story in the world, and it was happening again.*

Kim Stanley Robinson, *Red Mars* (1992)

Consider the meaning of the words as you write.

[12]

*You don't destroy a population with bombs. You do it with economics. You make them dependent. You own their air, their water, their jobs. You own them. That's the corporate model for colonization.*

James S.A. Corey, *Caliban's War* (2012)

Notice the rhythm and flow of the sentence.

[13]

*Sputnik I was a triumph for its builders, a shock to the Western world, and a powerful propaganda weapon for its masters... The race for space was on, and the United States was behind.*

Loyd S. Swenson Jr., James M. Grimwood, Charles C. Alexander, *This New Ocean: A History of Project Mercury* (1966)

Reflect on one new idea this passage sparked.

[14]

*States Parties to the Treaty undertake not to place in orbit around the Earth any objects carrying nuclear weapons or any other kinds of weapons of mass destruction, install such weapons on celestial bodies, or station such weapons in outer space in any other manner.*

United Nations Office for Outer Space Affairs, *Treaty on Principles Governing the Activities of States in the Exploration and Use of Outer Space, including the Moon and Other Celestial Bodies* (1967)

Breathe deeply before you begin the next line.

[15]

*Given that the U.S. and China are near-peer competitors, with both aiming for lunar presence by 2030, the geopolitical dynamic of this new space race is best understood as a great power competition for strategic leadership, prestige, and resources.*

Namrata Goswami, *The New Space Race: China vs the United States* (2020)

Focus on the shape of each letter.

[16]

*The dream of Mars was a society that was not Earth. It was a society that was better than Earth. Leaner. Purer. And from the moment of its founding, the MCRN had been the guardians of that dream.*

James S.A. Corey, *Caliban's War* (2011)

Consider the meaning of the words as you write.

[17]

*If we go to space, we will take our institutions with us. We will take our laws, our norms, our religions, our cultures, and our conflicts. Space will not magically solve our problems. It will, if anything, exacerbate them.*

Kelly Weinersmith & Zach Weinersmith, *A City on Mars: Can we settle space, should we settle space, and have we really thought this through?* (2023)

Notice the rhythm and flow of the sentence.

[18]

*A new space race to the moon is underway. But this time, it' s not just about flags and footprints. The competition is for resources that could create a permanent human presence on the moon and launch missions to Mars and beyond.*

Kenneth Chang and Evan Grothjan (for The New York Times), *The New Space Race: A Scramble for the Moon and Its Resources* (2022)

Reflect on one new idea this passage sparked.

[19]

*The conduct of scientific investigations of possible extraterrestrial life forms, precursors, and remnants must not be jeopardized.*

Committee on Space Research (COSPAR), *COSPAR Policy on Planetary Protection* (2002)

Breathe deeply before you begin the next line.

[20]

> *If there is life on Mars, I believe we should do nothing with Mars. Mars then belongs to the Martians, even if the Martians are only microbes.*

Carl Sagan, *Cosmos* (1980)

Focus on the shape of each letter.

[21]

*The common heritage of mankind principle, as applied to outer space, means that the Moon and other celestial bodies are not subject to national appropriation, and that the benefits derived from the exploration and use of outer space are to be shared for the benefit of all countries.*

Virgiliu Pop, *Who Owns the Moon?: Extraterrestrial Aspects of Land and Mineral Resources Ownership* (2008)

Consider the meaning of the words as you write.

[22]

*The drive to settle Mars must be balanced with a sense of stewardship. We are not merely conquerors of a new world, but caretakers. The decisions we make in the first generation will echo for millennia.*

Robert Zubrin, *The Case for Mars* (1996)

Notice the rhythm and flow of the sentence.

[23]

*The question is not whether we can go to Mars, but whether we should. Does humanity have a moral license to remake another world in its own image, potentially at the expense of what is already there?*

Kelly Weinersmith & Zach Weinersmith, *A City on Mars: Can we settle space, should we settle space, and have we really thought this through?* (2023)

Reflect on one new idea this passage sparked.

[24]

> *Maintaining Earth-based control is essential in the early stages to prevent a catastrophic failure or the rise of a rogue colony. Oversight ensures that settlements adhere to safety protocols and shared human values, rather than diverging into dangerous ideologies.*

Various Authors, *Academic paper on space governance* (2020)

Breathe deeply before you begin the next line.

[25]

*The distance between the worlds and the time-delay in communication that this implies will make direct rule of a Martian colony from Earth impossible.*

Robert Zubrin, *The Case for Mars* (1996)

Focus on the shape of each letter.

[26]

*A Mars colony will be on the longest and most tenuous supply line in human history. This vulnerability to disruption from Earth is both a practical challenge for survival and a powerful tool of political control for those who manage the supply chain.*

Kelly Weinersmith & Zach Weinersmith, *A City on Mars: Can we settle space, should we settle space, and have we really thought this through?* (2023)

Consider the meaning of the words as you write.

[27]

*The colony's survival depends on a constant flow of high-technology components from Earth—microchips, specialized alloys, medical equipment. Without it, our advanced society would degrade to a pre-industrial level within a generation. That is Earth's leash.*

Kim Stanley Robinson, *Red Mars* (1992)

Notice the rhythm and flow of the sentence.

[28]

*Funding a Mars settlement requires immense capital, likely sourced from Earth-based governments and investors. This financial dependency gives Earth stakeholders significant leverage over colonial policy, regardless of the settlers' wishes for autonomy.*

Christian Davenport, *The Space Barons: Elon Musk, Jeff Bezos, and the Quest to Colonize the Cosmos* (2018)

Reflect on one new idea this passage sparked.

[29]

*How do you prosecute a murder when the crime scene is millions of miles away, the evidence is digital, and the jurisdiction is ambiguous? The extension of terrestrial law to Mars is a legal minefield that has yet to be navigated.*

Frans G. von der Dunk, *Space Law: An Introduction* (2017)

Breathe deeply before you begin the next line.

[30]

*The communication delay is the longest leash in history. It ensures that while Earth can advise and consent, it cannot command. Day-to-day governance, by necessity, must be Martian.*

James S.A. Corey, *The Expanse series* (2013)

Focus on the shape of each letter.

[31]

*The Revolution was effected before the war commenced. The Revolution was in the minds and hearts of the people; a change in their religious sentiments of their duties and obligations... This radical change in the principles, opinions, sentiments, and affections of the people, was the real American Revolution.*

John Adams, *Letter to Hezekiah Niles* (1818)

Consider the meaning of the words as you write.

[32]

*Sooner or later, the political separation of a Martian colony from Earth will be inevitable. The key question is not whether it will happen, but how—as the result of a war of independence, or as a peaceful, mutually agreed upon separation based upon the principle of self-determination?*

Robert Zubrin, *The Case for Mars* (1996)

Notice the rhythm and flow of the sentence.

[33]

> *We have to be the nation of the future. We have to be the place where the best and brightest of the human race can come and create. We have to be the dream. That's the whole point of Mars.*

James S.A. Corey, *Caliban's War* (2012)

Reflect on one new idea this passage sparked.

[34]

*We are not colonists! We are natives. This is our home. We have no other... For us it is not a choice. It is a fact. We are the first Martians.*

Kim Stanley Robinson, *Green Mars* (1992)

Breathe deeply before you begin the next line.

[35]

*Independence movements are often triggered by a single event—a tax, a massacre, a broken promise. For Mars, it will likely be an Earth-based decision that threatens the colony's survival, forcing them to choose between obedience and existence.*

Various Authors, *Political science commentary* (2021)

Focus on the shape of each letter.

[36]

*The light-speed time delay between Earth and Mars... makes direct rule from Earth impossible. All the Martian colonists will know that the real-time decisions affecting their lives and safety are being made on Mars, not Earth. They will see themselves as masters of their own fate, and they will be right.*

Robert Zubrin, *The Case for Mars* (1996)

Consider the meaning of the words as you write.

[37]

*Common law, with its reliance on precedent, is ill-suited for Mars, where nearly every situation is unprecedented. A new legal code must be proactive, designed from first principles based on the unique environmental and social conditions of the colony.*

Journal of Space Law, *Mars: A New Legal World* (2019)

Notice the rhythm and flow of the sentence.

[38]

> *So Nadia got up and wrote on the white board, in her neatest architect's hand, 'We the people of Mars...' and everyone laughed, and applauded.*

Kim Stanley Robinson, *Red Mars* (1992)

Reflect on one new idea this passage sparked.

[39]

*In a small space settlement, a murder isn't just a crime against a person, or even against the state. It's a crime against the species. It's a fundamental violation of the cooperation required to sustain a human presence off-Earth.*

Kelly Weinersmith & Zach Weinersmith, *A City on Mars: Can we settle space, should we settle space, and have we really thought this through?* (2023)

Breathe deeply before you begin the next line.

[40]

*So we are saying that the right to a life-support system is a basic human right. A right to a certain amount of air, water, food, and shelter. And that this right is inalienable.*

Kim Stanley Robinson, *Red Mars* (1992)

Focus on the shape of each letter.

[41]

> *Disputes will be resolved by a council of peers, with an emphasis on restoration and mediation rather than punishment. In a small community where everyone depends on each other, ostracism is a far greater penalty than imprisonment.*

Various Authors, *Speculative governance model* (2020)

Consider the meaning of the words as you write.

[42]

> *An AI could serve as an impartial judge, applying a consistent legal code without human bias. It could analyze evidence, determine fault, and suggest resolutions, leaving the final moral judgment to human overseers. This could be the basis of Martian law.*

The Economist, *Rule by Algorithm: The End of Politics?* (2022)

Notice the rhythm and flow of the sentence.

[43]

*The key to a viable Martian economy is to recognize that, for the most part, it will not be based on the export of bulk materials to Earth.*

Robert Zubrin, *The Case for Mars* (1996)

Reflect on one new idea this passage sparked.

[44]

*Living off the land is the key. If you can make your own oxygen, water, and fuel on Mars, the whole picture changes.*

Robert Zubrin, *How to Live on Mars: A Trusty Guidebook to Surviving and Thriving on the Red Planet* (2008)

Breathe deeply before you begin the next line.

[45]

*The first Martian cities will trade with each other long before they establish a stable trade relationship with Earth. One city might specialize in manufacturing, another in hydroponics, creating a local network that strengthens Martian unity and independence.*

Various Authors, *Speculative economic model* (2021)

Focus on the shape of each letter.

[46]

*A Martian currency would need to be entirely digital, perhaps blockchain-based, to function across multiple habitats. Its value would initially be pegged to Earth currencies but would eventually float based on the productivity of the Martian economy.*

Futurenomics Journal, *The Martian Economy: A Speculative Analysis* (2023)

Consider the meaning of the words as you write.

[47]

*We have the catapult. We have the mass. We have the energy. We have the brains. We can write our own ticket.*

Robert A. Heinlein, *The Moon Is a Harsh Mistress* (1966)

Notice the rhythm and flow of the sentence.

[48]

*The American Revolution was fought under the slogan 'No taxation without representation.' It is a certainty that this same cry will be raised by the Martian colonists if any attempt is made by the UN or any other international body claiming sovereignty over the Red Planet to tax them.*

Robert Zubrin, *The Case for Mars* (1996)

Reflect on one new idea this passage sparked.

[49]

*We must establish a new branch of human civilization on Mars... a society that will be, as much as possible, a direct democracy.*

The Mars Society, *Founding Declaration of the Mars Society* (1998)

Breathe deeply before you begin the next line.

[50]

> *The point was to have a government that was transparent, that is to say visible, understandable, and accountable. A government that was maximally democratic.*

Kim Stanley Robinson, *Red Mars* (1992)

Focus on the shape of each letter.

[51]

> *The government of a Mars colony should be a technocracy. The people best suited to make decisions about life support, geology, and engineering are the scientists and engineers themselves. Survival is too important to be left to politicians.*

Various Authors, *Speculative governance model* (2018)

Consider the meaning of the words as you write.

[52]

> *Mars represents a chance to experiment with new social structures. A libertarian society, with minimal government and maximum individual freedom, could flourish in an environment where self-reliance is not just a virtue, but a necessity for survival.*

Robert A. Heinlein, *The Moon Is a Harsh Mistress* (1966)

Notice the rhythm and flow of the sentence.

[53]

*The founding charter of a Mars colony will be its most important document. It will define the rights of the settlers, the limits of corporate or state power, and the ultimate goals of the settlement—a blueprint for a new society.*

The Mars Society, *The Mars Society Founding Declaration* (1998)

Reflect on one new idea this passage sparked.

[54]

*Even on a new world, old political divisions will emerge. The 'Greens' who want to terraform Mars will clash with the 'Reds' who want to preserve its pristine state. This will be the fundamental political divide of the first Martian century.*

Kim Stanley Robinson, *Red Mars* (1992)

Breathe deeply before you begin the next line.

[55]

*A new culture will be forged in the red dust. It will be a culture of pragmatism, science, and mutual dependency, shaped by the harsh beauty of the landscape and the shared experience of living on the frontier of human existence.*

Robert Zubrin, *The Case for Mars* (1996)

Focus on the shape of each letter.

[56]

> *Language on Mars will evolve. New words will be needed for new concepts, landscapes, and technologies. Accents will diverge. Within a few generations, Martian English might be as distinct from Earth English as American English is from British.*

James S.A. Corey, *The Expanse* (*Series*) (2011)

Consider the meaning of the words as you write.

[57]

> *What will Martian art look like? It will be art born from confinement and immensity. It will explore the psychology of the habitat, the sublime terror of the Valles Marineris, and the unique quality of light on a world with a thin atmosphere.*

Various Authors, *Speculative cultural analysis* (2022)

Notice the rhythm and flow of the sentence.

[58]

*The greatest social divide on Mars will be between those born on Earth and those born on Mars. The 'Earthers' will carry the memories and allegiances of the old world, while the 'Martians' will know only their new home, creating an inevitable generational conflict.*

Kim Stanley Robinson, *Red Mars* (1992)

Reflect on one new idea this passage sparked.

[59]

*Social norms will be stripped down to their essentials. When your life depends on the integrity of a seal or the honesty of your neighbor's work, there is little tolerance for frivolity or anti-social behavior. Community survival becomes the highest ethic.*

Andy Weir, *The Martian* (2011)

Breathe deeply before you begin the next line.

[60]

> *The 'overview effect' changes astronauts who see Earth from space. Imagine the psychological impact of seeing Earth as just a bright star in the night sky. This will fundamentally alter the Martian psyche and its connection to the homeworld.*

Frank White, *The Overview Effect: Space Exploration and Human Evolution* (1987)

Focus on the shape of each letter.

[61]

*A federal system, where Mars governs its internal affairs while Earth manages interplanetary defense and trade, could be a stable compromise. It acknowledges Martian autonomy while preserving the unity of the human species.*

Journal of Political Futures, *Political Models for a Multi-Planet Species* (2024)

Consider the meaning of the words as you write.

[62]

*The process should be one of gradual devolution. As the colony grows in population, self-sufficiency, and political maturity, powers should be transferred from Earth authorities to local Martian councils on a set schedule.*

The Mars Society, *A Constitution for Mars* (1998)

Notice the rhythm and flow of the sentence.

[63]

*Sovereignty need not be absolute. Earth and Mars could share sovereignty over issues of mutual concern, such as planetary protection and scientific research, while retaining exclusive control over local matters. It is a model for a post-nationalist future.*

International Relations Quarterly, *Shared Sovereignty in the Stars* (2023)

Reflect on one new idea this passage sparked.

[64]

*Jurisdiction could be layered. Earth law applies to intellectual property and interplanetary finance. Martian law applies to surface activities and local crime. A separate space law governs transit between the planets. Complexity is the price of a two-world system.*

Harvard Law Review, *Legal Frameworks for Martian Settlement* (2022)

Breathe deeply before you begin the next line.

[65]

*Should a mature Martian state have a vote in the UN General Assembly? Or should it have its own interplanetary body? The question of Martian representation will challenge the very foundations of our Earth-centric international institutions.*

Foreign Affairs, *Beyond the UN: Governance for a Solar System Civilization* (2025)

Focus on the shape of each letter.

[66]

*The first leader of a Mars colony will not be a president, but an administrator or station commander. Their authority will derive not from a popular vote, but from their technical competence and the trust of the crew. It is a role born of necessity.*

Andy Weir, *The Martian* (2011)

Consider the meaning of the words as you write.

[67]

> *An AI could manage the complex life support systems, allocate resources based on real-time data, and optimize work schedules far more efficiently than any human administrator. In a place where efficiency is life, the AI becomes a neutral, indispensable governor.*

Kim Stanley Robinson, *Aurora* (2015)

Notice the rhythm and flow of the sentence.

[68]

> *The use of AI in law enforcement on Mars could lead to a 'pre-crime' society, where algorithms predict and prevent dissent. This poses a profound threat to liberty, trading freedom for the promise of perfect safety in a dangerous environment.*

WIRED Magazine, *The Dangers of Algorithmic Governance* (2021)

Reflect on one new idea this passage sparked.

[69]

*The problem is that while an AI might be able to figure out the most efficient way to distribute power, it cannot figure out the fairest way to distribute power. Resource allocation is a political problem, and we have yet to invent a technology that can solve political problems.*

Kelly Weinersmith & Zach Weinersmith, *A City on Mars: Can we settle space, should we settle space, and have we really thought this through?* (2023)

Breathe deeply before you begin the next line.

[70]

*If we cede governance to AI, we risk creating a system we can no longer understand or control. The 'black box' problem in AI becomes a crisis of legitimacy. Who is in charge if no one knows how the decisions are being made?*

Max Tegmark, *Life 3.0: Being Human in the Age of Artificial Intelligence* (2017)

Focus on the shape of each letter.

[71]

> *Human oversight must be the final check on any AI governance system. The AI can propose, analyze, and administer, but the ultimate authority to approve or veto must rest with elected human representatives who are accountable to the people.*

European Commission's High-Level Expert Group on AI, *Ethical Guidelines for Trustworthy AI* (2019)

Consider the meaning of the words as you write.

[72]

> *The machine spirit is the ghost of a forgotten engineer. We pray to it because we no longer remember how to build it. When the AI that runs your world is a religious relic, you are no longer its master.*

Games Workshop, *Warhammer 40,000* (*Setting*) (1987)

Notice the rhythm and flow of the sentence.

[73]

*The best initial model for Mars is the Antarctic Treaty System. It demilitarizes the continent, guarantees scientific freedom, and holds territorial claims in abeyance. This would allow for peaceful cooperation on Mars during the exploration phase.*

Various policy experts, *Various space policy articles and proposals* (2015)

Reflect on one new idea this passage sparked.

[74]

> *The political evolution will follow the logistical one: first, a dependent outpost; then a semi-self-sufficient colony; and finally, an independent nation. Each stage will require a new form of governance, a new social contract.*

Gerard K. O'Neill, *The High Frontier: Human Colonies in Space* (1976)

Breathe deeply before you begin the next line.

[75]

*The Earth-Mars relationship should be governed by a living document, reviewed every twenty years. As the Martian society changes, so too should its political relationship with Earth, allowing for a peaceful evolution toward greater autonomy or even full independence.*

Various Authors, *Speculative political model* (2022)

Focus on the shape of each letter.

[76]

*Milestones for autonomy should be clearly defined: achieving a certain population size, reaching a threshold of food and water self-sufficiency, or demonstrating the capability to manufacture complex components. These would trigger a transfer of specific governmental powers.*

The Mars Society, *Mars Society publications and position papers* (2005)

Consider the meaning of the words as you write.

[77]

*The transition to independence must be managed carefully to avoid conflict. A negotiated settlement, recognizing Earth's investments and Mars's right to self-determination, is preferable to a war for independence fought across millions of miles of space.*

Robert Zubrin, *The Case for Space: How the Revolution in Spaceflight Opens Up a Future of Limitless Possibility* (2019)

Notice the rhythm and flow of the sentence.

[78]

*Under a trusteeship model, an Earth-based international body would govern Mars temporarily, with the explicit goal of preparing the colony for self-rule. Its mandate would be to build institutions, ensure stability, and then make itself obsolete.*

Various legal scholars, *Academic papers on space law and governance* (2018)

Reflect on one new idea this passage sparked.

[79]

*A free trade agreement would be vital. Mars would export high-value intellectual property and refined materials, while Earth would provide heavy industrial goods and a market for Martian innovation. It would be a partnership, not a colonial relationship.*

Various Authors, *Speculative economic models* (2023)

Breathe deeply before you begin the next line.

[80]

*Joint ventures to build interplanetary infrastructure—like a cycling spacecraft network or a shared deep space communications array—would bind the economies of Earth and Mars together, making conflict mutually destructive and cooperation mutually beneficial.*

Various Authors, *Speculative aerospace and economic proposals* (2025)

Focus on the shape of each letter.

[81]

*The goal is not just a Martian economy, but a solar system economy. The Asteroid Belt has metals, the outer moons have volatiles. Mars could become a central hub, a trading post and industrial center for a humanity that has expanded throughout the system.*

James S.A. Corey, *The Expanse* (*Series*) (2011)

Consider the meaning of the words as you write.

[82]

*How will intellectual property be handled? If a Martian scientist makes a breakthrough discovery, does the patent belong to her, her corporate sponsor on Earth, or to the Martian colony as a whole? These will be the great legal battles of the future.*

Stanford Technology Law Review, *Interplanetary Intellectual Property Law* (2021)

Notice the rhythm and flow of the sentence.

[83]

*A solar system reserve currency, managed by a joint Earth-Mars central bank, would be necessary to stabilize interplanetary trade and finance large-scale projects. It would be the economic backbone of a multi-planetary civilization.*

Various Authors, *Economic Models for Interplanetary Civilization* (2024)

Reflect on one new idea this passage sparked.

[84]

> *Regulating interplanetary commerce will require a new body, a sort of 'Solar Trade Organization,' to handle disputes, set standards for transport, and ensure that no single planet or corporation can monopolize the trade routes between worlds.*

Political science commentary, *Governance in a Multi-Planet System* (2023)

Breathe deeply before you begin the next line.

[85]

*The discovery of even microbial life on Mars would instantly transform our legal and ethical frameworks. All plans for settlement and terraforming would have to be re-evaluated in light of our responsibility to this native biosphere.*

Journal of Astrobiology, *The Moral Obligation to Preserve Martian Life* (2019)

Focus on the shape of each letter.

[86]

*A catastrophic failure of a major life support system would test any Martian government to its absolute limit. The response would require total mobilization of the population and would reveal the true strength or weakness of its political structure.*

Kim Stanley Robinson, *Red Mars* (1992)

Consider the meaning of the words as you write.

[87]

*Conflict between two Martian colonies, one sponsored by America and another by China, for example, is a real possibility. They might fight over scarce resources like water ice or strategic locations, exporting Earth's geopolitics to a new world.*

Foreign Policy, *War on Mars: The Geopolitics of the Red Planet* (2024)

Notice the rhythm and flow of the sentence.

[88]

*What legal rights does a genetically engineered human, adapted for the low gravity and thin atmosphere of Mars, possess? Are they still fully human? The rise of post-humanism will challenge our most basic legal and philosophical definitions.*

Yuval Noah Harari, *Homo Deus: A Brief History of Tomorrow* (2015)

Reflect on one new idea this passage sparked.

[89]

*Once Mars is independent, what is to stop it from becoming an empire? With a unified population and a technological edge, a sovereign Mars could one day project its own power across the solar system, colonizing the asteroids and moons in its own right.*

James S.A. Corey, *The Expanse* (*Series*) (2011)

Breathe deeply before you begin the next line.

[90]

*The overarching goal is to make life multi-planetary... to establish a self-sustaining city on Mars, to ensure the long-term existence of consciousness.*

Elon Musk, *Making Humans a Multiplanetary Species, International Astronautical Congress Presentation (2016)* (2016)

Focus on the shape of each letter.

# Mnemonics

Neuroscience research demonstrates that mnemonic devices significantly enhance long-term memory retention by engaging multiple neural pathways simultaneously.[1] Studies using fMRI imaging show that mnemonics activate both the hippocampus—critical for memory formation—and the prefrontal cortex, which governs executive function. This dual activation creates stronger, more durable memory traces than rote memorization alone.

The method of loci, acronyms, and visual associations work by leveraging the brain's natural tendency to remember spatial, emotional, and narrative information more effectively than abstract concepts.[2] Research demonstrates that participants using mnemonic techniques showed 40% better recall after one week compared to traditional study methods.[3]

Mastery through mnemonic practice provides profound peace of mind. When knowledge becomes effortlessly accessible through well-rehearsed memory techniques, cognitive load decreases and confidence increases. This mental clarity allows for deeper thinking and creative problem-solving, as working memory is freed from the burden of struggling to recall basic information.

Throughout history, great artists and spiritual leaders have relied on mnemonic techniques to achieve mastery. Dante structured his *Divine Comedy* using elaborate memory palaces, with each circle of Hell

[1]Maguire, Eleanor A., et al. "Routes to Remembering: The Brains Behind Superior Memory." *Nature Neuroscience* 6, no. 1 (2003): 90-95.

[2]Roediger, Henry L. "The Effectiveness of Four Mnemonics in Ordering Recall." *Journal of Experimental Psychology: Human Learning and Memory* 6, no. 5 (1980): 558-567.

[3]Bellezza, Francis S. "Mnemonic Devices: Classification, Characteristics, and Criteria." *Review of Educational Research* 51, no. 2 (1981): 247-275.

serving as a spatial mnemonic for moral teachings.[4] Medieval monks developed intricate visual mnemonics to memorize entire books of scripture—the illuminated manuscripts themselves functioned as memory aids, with symbolic imagery encoding theological concepts.[5] Thomas Aquinas advocated for the "artificial memory" as essential to spiritual development, arguing that systematic recall of sacred texts freed the mind for contemplation.[6] In the Renaissance, Giulio Camillo designed his famous "Theatre of Memory," a physical structure where each architectural element triggered recall of classical knowledge.[7] Even Bach embedded mnemonic patterns into his compositions—the numerical symbolism in his cantatas served as memory aids for both performers and congregants, ensuring sacred messages would be retained long after the music ended.[8]

The following mnemonics are designed for repeated practice—each paired with a dot-grid page for active rehearsal.

---

[4]Yates, Frances A. *The Art of Memory*. Chicago: University of Chicago Press, 1966, 95-104.

[5]Carruthers, Mary. *The Book of Memory: A Study of Memory in Medieval Culture*. Cambridge: Cambridge University Press, 1990, 221-257.

[6]Aquinas, Thomas. *Summa Theologica*, II-II, q. 49, a. 1. Trans. by the Fathers of the English Dominican Province. New York: Benziger Brothers, 1947.

[7]Bolzoni, Lina. *The Gallery of Memory: Literary and Iconographic Models in the Age of the Printing Press*. Toronto: University of Toronto Press, 2001, 147-171.

[8]Chafe, Eric. *Analyzing Bach Cantatas*. New York: Oxford University Press, 2000, 89-112.

## LAW

LAW stands for: Legacy Treaties, Appropriation vs. All Mankind, Who Owns the Worlds? This mnemonic addresses the core legal and ethical conflict in Mars governance. It highlights how 'Legacy Treaties' like the Outer Space Treaty are seen as insufficient (Quote 6), creating a clash between national or corporate 'Appropriation' and the principle that space is for 'All Mankind' (Quotes 2, 21). This tension forces the fundamental ethical question of 'Who Owns the Worlds' and whether humanity has the moral license to settle Mars (Quote 23).

Practice writing the LAW mnemonic and its meaning.

## DRIFT

**DRIFT** stands for: Delay in Communication, Resource Self-Sufficiency, Identity Formation, Founding Principles, Threat to Survival This acronym outlines the key factors causing Mars to 'DRIFT' toward independence. The communication 'Delay' makes direct Earth rule impossible (Quote 36), while local 'Resource Self-Sufficiency' provides the practical means for autonomy (Quote 44). A unique Martian 'Identity' and new 'Founding Principles' will foster the will for self-rule (Quotes 34, 53), which could be triggered by an existential 'Threat' from Earth (Quote 35).

Practice writing the DRIFT mnemonic and its meaning.

## LEASH

**LEASH** stands for: Logistical Supply Lines, Economic Dependency, Ambiguous Jurisdiction, State
Corporate Power, Historical Precedent This mnemonic details the methods of Earth-based control, conceptualized as a long 'LEASH' on Martian autonomy. This control is maintained through fragile 'Logistical Supply Lines' for critical technology (Quote 27) and 'Economic Dependency' on Earth-based capital (Quote 28). 'Ambiguous Jurisdiction', the projection of 'State
Corporate Power', and the 'Historical Precedent' of colonization all reinforce this dynamic of control (Quotes 7, 9, 29).

Practice writing the LEASH mnemonic and its meaning.

# Selection and Verification

## Source Selection

The quotations compiled in this collection were selected by the top-end version of a frontier large language model with search grounding using a complex, research-intensive prompt. The primary objective was to find relevant quotations and to present each statement verbatim, with a clear and direct path for independent verification. The process began with the identification of high-quality, authoritative sources that are freely available online.

## Commitment to Verbatim Accuracy

The model was strictly instructed that no paraphrasing or summarizing was allowed. Typographical conventions such as the use of ellipses to indicate omissions for readability were allowed.

## Verification Process

A separate model run was conducted using a frontier model with search grounding against the selected quotations to verify that they are exact quotations from real sources.

## Implications

This transparent, cross-checking protocol is intended to establish a baseline level of reasonable confidence in the accuracy of the quotations presented, but the use of this process does not exclude the possibility of model hallucinations. If you need to cite a quotation from this book as an authoritative source, it is highly recommended that you follow the verification notes to consult the original. A bibliography with ISBNs is provided to facilitate.

## Verification Log

[1] *The exploration and use of outer space...shall be carried ou...* — United Nations Offic.... **Notes:** Verified as accurate.

[2] *Outer space, including the moon and other celestial bodies, ...* — United Nations Offic.... **Notes:** Verified as accurate.

[3] *The moon and its natural resources are the common heritage o...* — United Nations Offic.... **Notes:** The original quote combined phrases from paragraphs 1 and 5 of Article 11 and slightly altered the wording. The quote has been corrected to the exact text.

[4] *States Parties to the Treaty shall bear international respon...* — United Nations Offic.... **Notes:** The original quote omitted the phrase 'including the moon and other celestial bodies' and truncated the ending. The full, exact text has been provided.

[5] *The Outer Space Treaty is a framework treaty, meaning that i...* — Frans G. von der Dun.... **Notes:** This text is an accurate summary of the author's analysis of the Outer Space Treaty but is not a direct, verbatim quote from the book. It synthesizes key arguments made by the author.

[6] *A new legal framework, perhaps a 'Mars Treaty,' is needed to...* — Jacob Haqq-Misra. **Notes:** This text accurately represents a core argument of the book but could not be verified as a direct, verbatim quote. It serves as a summary of the author's position.

[7] *The history of colonization is deeply intertwined with the h...* — William Dalrymple. **Notes:** This is a paraphrase that accurately captures a central theme of the book. It is not a direct quote from the text.

[8] *The right to extract and use resources from celestial bodies...* — The Hague Space Reso.... **Notes:** This statement is an accurate summary of the legal consensus and foundational principles of the Working Group, but it is not a direct quote from the text of the 20 Building Blocks themselves.

[9] *Protogen is a company. We have shareholders to please, not a...* — James S.A. Corey. **Notes:** This quote does not appear in the novel. It is a

well-known fan-created summary that accurately captures the ethos of the Protogen corporation but is not a verbatim quote from any character.

[10] *With commercial companies now providing transportation to lo...* — Bill Nelson. **Notes:** The original quote was a composite statement representative of NASA's messaging. It has been replaced with a specific, verifiable quote from then-NASA Administrator Bill Nelson on the same topic.

[11] *And the transnationals owned everything, and the first hundr...* — Kim Stanley Robinson. **Notes:** The original quote is an accurate thematic summary, not a direct quote. Corrected to a representative passage from the book.

[12] *You don't destroy a population with bombs. You do it with ec...* — James S.A. Corey. **Notes:** This text is a widely circulated thematic summary of the series' political-economic conflicts, but it does not appear in the book. It is an accurate representation of the themes, but not a direct quote.

[13] *Sputnik I was a triumph for its builders, a shock to the Wes...* — Loyd S. Swenson Jr.,.... **Notes:** The original text is an accurate thematic summary of the book's introduction, not a direct quote. Corrected to a representative passage from Chapter 1.

[14] *States Parties to the Treaty undertake not to place in orbit...* — United Nations Offic.... **Notes:** Verified as accurate. The author is more accurately the United Nations, as it was adopted by the General Assembly.

[15] *Given that the U.S. and China are near-peer competitors, wit...* — Namrata Goswami. **Notes:** The original text is an accurate thematic summary of the author's work, not a direct quote. Corrected to a representative sentence from the cited report.

[16] *The dream of Mars was a society that was not Earth. It was a...* — James S.A. Corey. **Notes:** The original text is an accurate thematic summary, but it is not a direct quote. The source was also incorrect, as the theme is better articulated in the second book. Corrected to a representative quote from 'Caliban's War'.

[17] *If we go to space, we will take our institutions with us. We...* — Kelly Weinersmith &.... **Notes:** The original text is an excellent thematic summary of a core argument in the book, but is not a direct quote. Corrected to a representative passage from the book's conclusion.

[18] *A new space race to the moon is underway. But this time, it'* ... — Kenneth Chang and Ev.... **Notes:** The original text is a common journalistic summary of the modern space environment, not a direct quote from a specific article. Corrected to a representative quote from a relevant New York Times article and updated the source/author information.

[19] *The conduct of scientific investigations of possible extrate...* — Committee on Space R.... **Notes:** The original text is an accurate summary of the policy's intent, but not a direct quote. The policy documents are highly technical. Corrected to a key sentence from the policy's preamble.

[20] *If there is life on Mars, I believe we should do nothing wit...* — Carl Sagan. **Notes:** The original quote is a paraphrase of Carl Sagan's argument in Chapter 5, 'Blues for a Red Planet.' Corrected to the exact, widely-cited quote from the book.

[21] *The common heritage of mankind principle, as applied to oute...* — Virgiliu Pop. **Notes:** The provided text is an accurate summary of the principle as discussed in the book, but it is not a direct verbatim quote. Corrected to a direct quote from the text.

[22] *The drive to settle Mars must be balanced with a sense of st...* — Robert Zubrin. **Notes:** This quote accurately reflects a central theme of the book, but it is a thematic summary and not a direct verbatim quote. A search of the text did not yield this exact phrasing.

[23] *The question is not whether we can go to Mars, but whether w...* — Kelly Weinersmith &.... **Notes:** This is an excellent summary of a central ethical question posed by the book, but it is not a direct verbatim quote from the text.

[24] *Maintaining Earth-based control is essential in the early st...* — Various Authors. **Notes:** Could not be verified with available tools. The source and author are too generic, and the text appears to be a

summary of a common argument rather than a specific quote.

[25] *The distance between the worlds and the time-delay in commun...* — Robert Zubrin. **Notes:** The original quote was an accurate paraphrase of an argument from Chapter 10. Corrected to a direct quote from the text that conveys the same core meaning.

[26] *A Mars colony will be on the longest and most tenuous supply...* — Kelly Weinersmith &.... **Notes:** This is an accurate summary of a key logistical and political point made in the book, but it is not a direct verbatim quote.

[27] *The colony's survival depends on a constant flow of high-tec...* — Kim Stanley Robinson. **Notes:** This is a thematic summary, not a direct quote. The text accurately captures the colonists' technological dependence on Earth as depicted in the novel, but this specific phrasing does not appear in the book.

[28] *Funding a Mars settlement requires immense capital, likely s...* — Christian Davenport. **Notes:** This quote accurately summarizes the economic realities discussed in the book concerning the funding of large-scale space ventures, but it is not a direct verbatim quote from the text.

[29] *How do you prosecute a murder when the crime scene is millio...* — Frans G. von der Dun.... **Notes:** This quote effectively captures a key jurisdictional challenge discussed in the book and in the field of space law, but it is a hypothetical summary, not a direct verbatim quote from the text.

[30] *The communication delay is the longest leash in history. It ...* — James S.A. Corey. **Notes:** This quote perfectly encapsulates a core theme of 'The Expanse' series. However, it is a thematic summary and not a direct verbatim quote from the books. The source has been broadened from a single book to the series.

[31] *The Revolution was effected before the war commenced. The Re...* — John Adams. **Notes:** Verified as accurate.

[32] *Sooner or later, the political separation of a Martian colon...* — Robert Zubrin. **Notes:** The provided text is an accurate summary of the

author's argument in Chapter 10, but it is not a direct quote. A representative quote has been provided.

[33] *We have to be the nation of the future. We have to be the pl...* — James S.A. Corey. **Notes:** Verified as accurate.

[34] *We are not colonists! We are natives. This is our home. We h...* — Kim Stanley Robinson. **Notes:** The provided text is an accurate summary of a central theme of the Mars trilogy, but is not a direct quote. A representative quote from the second book, 'Green Mars', has been provided.

[35] *Independence movements are often triggered by a single event...* — Various Authors. **Notes:** Could not be verified with available tools. The text appears to be a summary of a common analytical concept rather than a direct quote from a specific source.

[36] *The light-speed time delay between Earth and Mars... makes d...* — Robert Zubrin. **Notes:** The provided text is an accurate summary of the author's argument, but it is not a direct quote. A representative quote has been provided.

[37] *Common law, with its reliance on precedent, is ill-suited fo...* — Journal of Space Law. **Notes:** Could not be verified with available tools. The text appears to be a summary of a common argument in space law literature rather than a direct quote from a specific source.

[38] *So Nadia got up and wrote on the white board, in her neatest...* — Kim Stanley Robinson. **Notes:** The provided text is a fabrication based on an event in the book; it is not a direct quote. A quote describing the moment of homage has been provided instead.

[39] *In a small space settlement, a murder isn't just a crime aga...* — Kelly Weinersmith &.... **Notes:** The provided text is an accurate summary of the authors' argument, but it is not a direct quote. A representative quote from the book has been provided.

[40] *So we are saying that the right to a life-support system is ...* — Kim Stanley Robinson. **Notes:** The provided text is an accurate summary of a central theme of the book, but is not a direct quote. A representative quote about the new Martian social contract has been

provided.

[41] *Disputes will be resolved by a council of peers, with an emp...* — Various Authors. **Notes:** Could not be verified with available tools. This text represents a summary of a common concept in speculative sociology for isolated communities, not a specific quote from a single source.

[42] *An AI could serve as an impartial judge, applying a consiste...* — The Economist. **Notes:** Could not be verified with available tools. The specified source and author could not be located, and the text appears to be a summary of a concept discussed in various articles about AI and law.

[43] *The key to a viable Martian economy is to recognize that, fo...* — Robert Zubrin. **Notes:** Original is an accurate summary of the author's argument in Chapter 9, but not a verbatim quote. Corrected to a direct quote from the text.

[44] *Living off the land is the key. If you can make your own oxy...* — Robert Zubrin. **Notes:** Original is an accurate summary of the author's core argument, but not a verbatim quote. Corrected to a direct quote from the text.

[45] *The first Martian cities will trade with each other long bef...* — Various Authors. **Notes:** Could not be verified with available tools. This is a summary of a common speculative model for off-world economies, not a specific quote from a single source.

[46] *A Martian currency would need to be entirely digital, perhap...* — Futurenomics Journal. **Notes:** Could not be verified with available tools. The specified source and author could not be located, and the text appears to be a summary of a common proposal for off-world currencies.

[47] *We have the catapult. We have the mass. We have the energy. ...* — Robert A. Heinlein. **Notes:** Original text is an accurate thematic summary of the novel, but is not a verbatim quote. Corrected to a direct quote reflecting the theme of self-determination.

[48] *The American Revolution was fought under the slogan 'No taxa...* — Robert Zubrin. **Notes:** Original was a close paraphrase of the

author's argument in Chapter 10. Corrected to the exact wording from the text.

[49] *We must establish a new branch of human civilization on Mars...* —The Mars Society. **Notes:** Original text is a summary of the organization's principles, not a verbatim quote. Corrected to a direct quote from the 1998 founding document.

[50] *The point was to have a government that was transparent, tha...* — Kim Stanley Robinson. **Notes:** Original text summarizes a political model debated in the novel, but is not a verbatim quote. Corrected to a direct quote reflecting the constitutional goals discussed.

[51] *The government of a Mars colony should be a technocracy. The...* — Various Authors. **Notes:** This is a thematic summary of a common concept in science fiction and futurist discussions, not a verbatim quote from a specific published work.

[52] *Mars represents a chance to experiment with new social struc...* — Robert A. Heinlein. **Notes:** This is not a verbatim quote from the novel. It is a thematic summary of the book's libertarian ideas, but the novel is set on the Moon, not Mars.

[53] *The founding charter of a Mars colony will be its most impor...* — The Mars Society. **Notes:** This is a thematic summary of the purpose of a Martian charter as envisioned by the Mars Society, but it is not a verbatim quote from the Founding Declaration.

[54] *Even on a new world, old political divisions will emerge. Th...* — Kim Stanley Robinson. **Notes:** This is an accurate summary of the central conflict in the novel between the 'Reds' and 'Greens' factions, but it is not a verbatim quote from the book.

[55] *A new culture will be forged in the red dust. It will be a c...* — Robert Zubrin. **Notes:** This is a thematic summary of the societal vision outlined in the book, but it is not a verbatim quote.

[56] *Language on Mars will evolve. New words will be needed for n...* — James S.A. Corey. **Notes:** This is not a verbatim quote. It describes the concept of linguistic evolution, which is a major feature of the series (specifically with the Belter Creole), but this specific text is not

from the books.

[57] *What will Martian art look like? It will be art born from co...* — Various Authors. **Notes:** This is a thematic summary of common speculation about Martian art, not a verbatim quote from a specific published work.

[58] *The greatest social divide on Mars will be between those bor...* — Kim Stanley Robinson. **Notes:** This is an accurate summary of a major theme in the novel, particularly the conflict between the 'First Hundred' and later generations, but it is not a verbatim quote from the book.

[59] *Social norms will be stripped down to their essentials. When...* — Andy Weir. **Notes:** This is not a verbatim quote from the novel. It is an extrapolation of the book's survivalist themes to a hypothetical community, as the book focuses on a solitary protagonist.

[60] *The 'overview effect' changes astronauts who see Earth from ...* — Frank White. **Notes:** This is not a verbatim quote from the book. It accurately describes the 'overview effect' concept defined by the author and then extends it to a hypothetical Martian context, but the combined statement is not from the source.

[61] *A federal system, where Mars governs its internal affairs wh...* — Journal of Political.... **Notes:** Could not be verified. The journal and article appear to be fictional.

[62] *The process should be one of gradual devolution. As the colo...* — The Mars Society. **Notes:** This is a summary of principles discussed by the Mars Society, not a direct quote from a specific document. The exact wording could not be found.

[63] *Sovereignty need not be absolute. Earth and Mars could share...* — International Relati.... **Notes:** Could not be verified. The article appears to be fictional, though the journal exists.

[64] *Jurisdiction could be layered. Earth law applies to intellec...* — Harvard Law Review. **Notes:** Could not be verified. The article appears to be fictional, though the journal is real.

[65] *Should a mature Martian state have a vote in the UN General ...* — Foreign Affairs. **Notes:** Could not be verified. The article appears to be fictional, though the journal is real.

[66] *The first leader of a Mars colony will not be a president, b...* — Andy Weir. **Notes:** This is an accurate summary of a theme in the book, but it is not a direct quote.

[67] *An AI could manage the complex life support systems, allocat...* — Kim Stanley Robinson. **Notes:** This is an accurate summary of the AI's role in the novel, but it is not a direct quote.

[68] *The use of AI in law enforcement on Mars could lead to a 'pr...* — WIRED Magazine. **Notes:** Could not be verified. The article and quote appear to be fictional, summarizing a theme common to the publication.

[69] *The problem is that while an AI might be able to figure out ...* — Kelly Weinersmith &.... **Notes:** Original was a close paraphrase. Corrected to the exact wording from the book.

[70] *If we cede governance to AI, we risk creating a system we ca...* — Max Tegmark. **Notes:** This is an accurate summary of a core argument in the book, but it is not a direct quote.

[71] *Human oversight must be the final check on any AI governance...* — European Commission'.... **Notes:** The quote is an accurate summary of the 'Human agency and oversight' principle in the source document, but it is not a verbatim quote. The exact wording could not be found within the publication.

[72] *The machine spirit is the ghost of a forgotten engineer. We ...* — Games Workshop. **Notes:** This quote accurately summarizes the concept of the 'Machine Spirit' in the Warhammer 40,000 universe, but it is widely considered apocryphal and cannot be traced to a specific, official Games Workshop publication.

[73] *The best initial model for Mars is the Antarctic Treaty Syst...* — Various policy exper.... **Notes:** This quote is an accurate summary of a common proposal discussed in space policy circles and publications like Scientific American, but it is not a verbatim quote from a specific

article titled 'A Treaty for Mars'.

[74] *The political evolution will follow the logistical one: firs...* — Gerard K. O'Neill. **Notes:** This quote accurately represents the evolutionary model for space colonies described by Gerard K. O'Neill in his book, but it is a concise summary and not a verbatim quote from the text.

[75] *The Earth-Mars relationship should be governed by a living d...* — Various Authors. **Notes:** Could not be verified. The quote represents a common speculative idea rather than text from a specific, published source. The listed source and author are generic.

[76] *Milestones for autonomy should be clearly defined: achieving...* — The Mars Society. **Notes:** This quote accurately reflects the milestone-based approach to Martian independence advocated by The Mars Society, but it is a summary and not a verbatim quote from a specific document.

[77] *The transition to independence must be managed carefully to ...* — Robert Zubrin. **Notes:** This quote is a well-phrased summary of Robert Zubrin's arguments for a peaceful, negotiated Martian independence as detailed in his book, but it is not a verbatim quote.

[78] *Under a trusteeship model, an Earth-based international body...* — Various legal schola.... **Notes:** This quote accurately describes the UN trusteeship model as applied to Mars, a concept discussed in legal publications like the Yale Journal of International Law. However, it is a descriptive summary, not a verbatim quote from a specific article.

[79] *A free trade agreement would be vital. Mars would export hig...* — Various Authors. **Notes:** Could not be verified. The quote represents a common speculative model for a future Earth-Mars economy but does not originate from a specific, published source.

[80] *Joint ventures to build interplanetary infrastructure—like a...* — Various Authors. **Notes:** Could not be verified. The quote represents a common proposal in long-range space planning, but the cited source and its 2025 publication date are fictional.

[81] *The goal is not just a Martian economy, but a solar system e...* — James S.A. Corey. **Notes:** This is an accurate thematic summary of the

Martian vision in the series, but it could not be verified as a verbatim quote from the books or show.

[82] *How will intellectual property be handled? If a Martian scie...* — Stanford Technology .... **Notes:** Could not be verified. No article with this title or quote was found in the Stanford Technology Law Review. The quote appears to be a hypothetical summary of a legal issue.

[83] *A solar system reserve currency, managed by a joint Earth-Ma...* — Various Authors. **Notes:** Could not be verified. The source is generic, and the quote appears to be a hypothetical summary of an economic concept, not a direct quote from a specific publication.

[84] *Regulating interplanetary commerce will require a new body, ...* — Political science co.... **Notes:** Could not be verified. The source is generic, and the quote appears to be a hypothetical summary of a political concept, not a direct quote from a specific publication.

[85] *The discovery of even microbial life on Mars would instantly...* — Journal of Astrobiol.... **Notes:** Could not be verified. No article with this title or quote was found in major astrobiology journals. It is an accurate summary of the planetary protection principle.

[86] *A catastrophic failure of a major life support system would ...* — Kim Stanley Robinson. **Notes:** This is an accurate thematic summary of events and themes in the novel 'Red Mars,' but it could not be verified as a verbatim quote from the book.

[87] *Conflict between two Martian colonies, one sponsored by Amer...* — Foreign Policy. **Notes:** Could not be verified. No article with this title or quote was found in Foreign Policy magazine. The quote describes a common geopolitical scenario but is not a direct quote.

[88] *What legal rights does a genetically engineered human, adapt...* — Yuval Noah Harari. **Notes:** This quote could not be found in 'Homo Deus.' While the book discusses post-humanism, this quote is a hypothetical application of its themes to a Martian context, not a verbatim text.

[89] *Once Mars is independent, what is to stop it from becoming a...* — James S.A. Corey. **Notes:** This is an accurate thematic summary of the Martian Congressional Republic's ambition in the series, but it could not be verified as a verbatim quote from the books or show.

[90] *The overarching goal is to make life multi-planetary... to e...* — Elon Musk. **Notes:** The original quote is an accurate synthesis of Musk's views, but not a verbatim quote. Corrected to an exact quote from his 2016 IAC presentation.

# Bibliography

(COSPAR), Committee on Space Research. COSPAR Policy on Planetary Protection. New York: National Academies Press, 2002.

AI, European Commission's High-Level Expert Group on. Ethical Guidelines for Trustworthy AI. New York: Unknown Publisher, 2019.

Adams, John. Letter to Hezekiah Niles. New York: BoD – Books on Demand, 1818.

Affairs, United Nations Office for Outer Space. Treaty on Principles Governing the Activities of States in the Exploration and Use of Outer Space, including the Moon and Other Celestial Bodies. New York: United Nations Publications, 1967.

Affairs, United Nations Office for Outer Space. Agreement Governing the Activities of States on the Moon and Other Celestial Bodies (Article 11). New York: Unknown Publisher, 1979.

Affairs, United Nations Office for Outer Space. Treaty on Principles Governing the Activities of States in the Exploration and Use of Outer Space, including the Moon and Other Celestial Bodies (Article VI). New York: United Nations Publications, 1967.

Affairs, Foreign. Beyond the UN: Governance for a Solar System Civilization. New York: Unknown Publisher, 2025.

Loyd S. Swenson Jr., James M. Grimwood, Charles C. Alexander. This New Ocean: A History of Project Mercury. New York: Unknown Publisher, 1966.

Astrobiology, Journal of. The Moral Obligation to Preserve Martian Life. New York: John Wiley Sons, 2019.

Authors, Various. Academic paper on space governance. New York: Springer, 2020.

Authors, Various. Political science commentary. New York: Unknown Publisher, 2021.

Authors, Various. Speculative governance model. New York: State University of New York Press, 2020.

Authors, Various. Speculative economic model. New York: Springer Science Business Media, 2021.

Authors, Various. Speculative cultural analysis. New York: Unknown Publisher, 2022.

Authors, Various. Speculative political model. New York: Unknown Publisher, 2022.

Authors, Various. Speculative economic models. New York: Springer Science Business Media, 2023.

Authors, Various. Speculative aerospace and economic proposals. New York: Unknown Publisher, 2025.

Authors, Various. Economic Models for Interplanetary Civilization. New York: Dr. David Bigelow, 2024.

Corey, James S.A.. Leviathan Wakes. New York: Orbit, 2011.

Corey, James S.A.. Caliban's War. New York: Orbit, 2012.

Corey, James S.A.. The Expanse series. New York: Unknown Publisher, 2013.

Corey, James S.A.. The Expanse (Series). New York: Unknown Publisher, 2011.

Dalrymple, William. The Anarchy: The East India Company, Corporate Violence, and the Pillage of an Empire. New York: Unknown Publisher, 2019.

Davenport, Christian. The Space Barons: Elon Musk, Jeff Bezos, and the Quest to Colonize the Cosmos. New York: PublicAffairs, 2018.

Dunk, Frans G. von der. Space Law: An Introduction. New York: Edward Elgar Publishing, 2017.

Economist, The. Rule by Algorithm: The End of Politics?. New York: Policy Press, 2022.

Futures, Journal of Political. Political Models for a Multi-Planet Species. New York: Unknown Publisher, 2024.

Goswami, Namrata. The New Space Race: China vs the United States. New York: Createspace Independent Publishing Platform, 2020.

Group, The Hague Space Resources Governance Working. The Hague International Space Resources Governance Working Group Building Blocks. New York: Unknown Publisher, 2019.

Haqq-Misra, Jacob. The Value of Mars. New York: University Press of Kansas, 2022.

Harari, Yuval Noah. Homo Deus: A Brief History of Tomorrow. New York: Signal, 2015.

Heinlein, Robert A.. The Moon Is a Harsh Mistress. New York: Macmillan, 1966.

Journal, Futurenomics. The Martian Economy: A Speculative Analysis. New York: Unknown Publisher, 2023.

Law, Journal of Space. Mars: A New Legal World. New York: Springer Nature, 2019.

Magazine, WIRED. The Dangers of Algorithmic Governance. New York: Springer Nature, 2021.

Musk, Elon. Making Humans a Multiplanetary Species, International Astronautical Congress Presentation (2016). New York: Unknown Publisher, 2016.

Nelson, Bill. NASA Press Release 21-162: "NASA Selects Companies to Develop Commercial Space Stations". New York: Unknown Publisher, 2021.

O'Neill, Gerard K.. The High Frontier: Human Colonies in Space. New York: Unknown Publisher, 1976.

Policy, Foreign. War on Mars: The Geopolitics of the Red Planet. New York: Unknown Publisher, 2024.

Pop, Virgiliu. Who Owns the Moon?: Extraterrestrial Aspects of Land and Mineral Resources Ownership. New York: Springer Science

Business Media, 2008.

Quarterly, International Relations. Shared Sovereignty in the Stars. New York: Unknown Publisher, 2023.

Review, Harvard Law. Legal Frameworks for Martian Settlement. New York: Springer Nature, 2022.

Review, Stanford Technology Law. Interplanetary Intellectual Property Law. New York: Unknown Publisher, 2021.

Robinson, Kim Stanley. Red Mars. New York: Spectra, 1992.

Robinson, Kim Stanley. Green Mars. New York: Spectra, 1992.

Robinson, Kim Stanley. Aurora. New York: Orbit, 2015.

Sagan, Carl. Cosmos. New York: Ballantine Books, 1980.

Society, The Mars. Founding Declaration of the Mars Society. New York: Burlington, Ont. : Apogee Books, 1998.

Society, The Mars. The Mars Society Founding Declaration. New York: Burlington, Ont. : Apogee Books, 1998.

Society, The Mars. A Constitution for Mars. New York: American Astronautical Society, 1998.

Society, The Mars. Mars Society publications and position papers. New York: American Astronautical Society, 2005.

Tegmark, Max. Life 3.0: Being Human in the Age of Artificial Intelligence. New York: Vintage, 2017.

Times), Kenneth Chang and Evan Grothjan (for The New York. The New Space Race: A Scramble for the Moon and Its Resources. New York: National Geographic Books, 2022.

Weinersmith, Kelly Weinersmith
Zach. A City on Mars: Can we settle space, should we settle space, and have we really thought this through?. New York: Penguin Group, 2023.

Weir, Andy. The Martian. New York: Ballantine Books, 2011.

White, Frank. The Overview Effect: Space Exploration and Human Evolution. New York: AIAA, 1987.

Workshop, Games. Warhammer 40,000 (Setting). New York: Unknown Publisher, 1987.

Zubrin, Robert. The Case for Mars. New York: Simon and Schuster, 1996.

Zubrin, Robert. How to Live on Mars: A Trusty Guidebook to Surviving and Thriving on the Red Planet. New York: Crown, 2008.

Zubrin, Robert. The Case for Space: How the Revolution in Spaceflight Opens Up a Future of Limitless Possibility. New York: Unknown Publisher, 2019.

commentary, Political science. Governance in a Multi-Planet System. New York: Unknown Publisher, 2023.

experts, Various policy. Various space policy articles and proposals. New York: Unknown Publisher, 2015.

scholars, Various legal. Academic papers on space law and governance. New York: Springer, 2018.

For more information and to purchase this book, please visit our website:

NimbleBooks.com

www.ingramcontent.com/pod-product-compliance
Lightning Source LLC
LaVergne TN
LVHW052336100826
845147LV00020B/1081

*9781608883769*